3-01
ch

THE STORY OF
WEIGHTS
AND
MEASURES

by Anita Ganeri

OXFORD UNIVERSITY PRESS

Published in the United States of America by
Oxford University Press, Inc.
198 Madison Avenue
New York, NY 10016

Oxford is a registered trademark of Oxford University Press, Inc.

Published in the United Kingdom by Evans Brothers Limited
2A Portman Mansions
Chiltern Street
London W1M 1LE

Printed in Hong Kong by Wing King Tong Co. Ltd

Library of Congress Cataloging-in-Publication Data
Ganeri, Anita.
    The story of weights and measures / by Anita Ganeri.
        p. cm. — (Signs of the times)
    Includes index.
    ISBN 0-19-521328-9
    1. Weights and measures — History — Juvenile literature.
[1. Weights and measures.] I. Title. II. Series: Ganeri, Anita, 1961-
Signs of the times
QC90.6.G36 1996
530.8'1'09 — dc20

96-32237
CIP
AC

Acknowledgments

Editor: Nicola Barber

Design: Neil Sayer

Illustrations: Hardlines

Production: Jenny Mulvanny

Acknowledgments

The author and publishers would like to thank the following for permission to reproduce photographs:
Front cover (top left) Catherine Pouedras, Science Photo Library, (top right) IPC Magazines, Robert Harding Picture Library, (center) Jean-Loup Charmet, Science Photo Library, (bottom left) Martyn F. Chillmaid, Robert Harding Picture Library
Back cover (left) Adrienne Hart-Davis, Science Photo Library, (right) Michael Dalton, Fundamental Photos, Science Photo Library
Title page Ronald Sheridan, Ancient Art & Architecture Collection
page 6 (top) British Museum, (bottom) David Vaughan, Science Photo Library page 7 (top) Ronald Sheridan, Ancient Art & Architecture Collection, (bottom) Dick Luria, Science Photo Library page 8 Ronald Sheridan, Ancient Art & Architecture Collection page 9 (top left) Last Resort Picture Library, (top right) J. Beecham, Ancient Art & Architecture Collection, (bottom) British Library, London, The Bridgeman Art Library page 10 (top) Robert Harding Picture Library, (bottom) British Library, London, The Bridgeman Art Library page 11 (top) Crown Copyright, Historic Royal Palaces, (bottom) ZEFA page 12 (top) Bibliothèque Nationale, Paris, The Bridgeman Art Library, (bottom) The Science Museum, Science & Society Picture Library page 13 (top) The Science Museum, Science & Society Picture Library, (middle) Tom McHugh, Science Photo Library, (bottom) IPC Magazines, Robert Harding Picture Library page 14 (top) NASA, Science Photo Library, (bottom) IPC Magazines, Robert Harding Picture Library page 15 (top) Garry Watson, Science Photo Library, (bottom) British Library, London, The Bridgeman Art Library page 16 Ronald Sheridan, Ancient Art & Architecture Collection page 17 (top) Tony Martin, Oxford Scientific Films, (left) Glyn Kirk, Action-Plus, (right) Ronald Sheridan, Ancient Art & Architecture Collection page 18 Luke Dodd, Science Photo Library page 19 (top) Robert Harding Picture Library, (middle) Tony Henshaw, Action-Plus, (bottom) Royal Geographical Society, London, The Bridgeman Art Library page 20 (top) James Stevenson, Science Photo Library, (bottom) Adam Hart-Davis, Science Photo Library page 21 (top left) Damien Lovegrove, Science Photo Library, (top right) Roger Ressmeyer, Starlight, Science Photo Library, (bottom) James Stevenson, Science Photo Library page 22 (top left) Science Photo Library, (middle) NOAO, Science Photo Library, (bottom right) Gavin Hellier, Robert Harding Picture Library, (bottom left) Robert Harding Picture Library page 23 (top) David Parker, Science Photo Library, (bottom) Walter Rawlings, Robert Harding Picture Library page 24 (top) E. Strouhal, Werner Forman Archive, (bottom) Bibliothèque Nationale, Paris, The Bridgeman Art Library page 25 (top) Ronald Sheridan, Ancient Art & Architecture Library, (middle) Werner Forman Archive, (bottom) Jean-Loup Charmet, Science Photo Library page 26 (top) Michael Dalton, Fundamental Photos, Science Photo Library, (bottom) Kim Westerskov, Oxford Scientific Films page 27 (top) Salter Weigh-Tronix Limited, A division of Staveley Industries plc., (bottom) Philippe Plailly, Eurelios, Science Photo Library.

CONTENTS

WEIGHTS AND MEASURES

How tall are you? What do you weigh? What size shoes do you wear? How far do you live from your school? How hot or cold is it today? In our daily lives, we use weights and measures to calculate height, weight, speed, distance, temperature, and area. Accurate ways of measuring are also essential in science, medicine, and the business world. This book tells the story of weights and measures from the earliest lumps of metal to the precision instruments of today.

Ancient Babylonian metal weights, in the shape of lions

BEFORE WEIGHTS

Early people had no set systems of weights or accurate measuring machines. They measured the size of things, such as spears or sheep, by comparing one with another to see which was bigger or smaller. They guessed how heavy things were by weighing them in their hands.

Today, scientists use sophisticated measuring instruments, such as this microwave distance measurer, to give extremely accurate results.

STANDARD MEASURES

In the past, people in different places used different sets of weights and measures. There were no fixed standards. This led to great confusion. Today, we use an internationally accepted system for most weights and measures (see page 13). But some measurements, such as shoe and clothes sizes, still vary from place to place. A British adult shoe size 3, for example, is an American size 4½ and a European size 36.

SIGNPOST

The first weights and measures were invented by the ancient Egyptians and Babylonians. They were used to calculate the weight of crops, to measure the size of fields, and for trading.

Ancient Egyptian farmers measure and record the size of a harvest.

LIGHTS, SPEED, ACTION!

The display in the cockpit of an airplane shows some modern electronic measuring instruments in action. There are gauges to show speed through the air, height above the ground (altitude), angle of climb, and a computerized runway diagram.

The computerized instrument panel in an airplane's cockpit

IN FACT...

Some units of measurement have unusual names. A catty is a Chinese unit, used mainly to measure tea. One catty equals about 1.3 pounds (.6 kilogram). A frail is a Spanish unit, used to measure raisins. A line (.024 inch/.6 millimeter) was once used in Europe to measure the size of buttons.

BODY MEASURES

Measurements are made in units, such as feet or centimeters for length, pounds or kilograms for weight, and degrees for temperature. The earliest system of units was based on parts of the human body, such as hands, arms, fingers, and feet. The only problem with this system was that the measurements varied, depending on how big or small the person was!

SIGNPOST

Farmers in ancient Egypt relied on the flooding of the Nile River every year to fertilize their land. But every year the floodwaters washed away the boundaries of the farmers' fields. The farmers used 12-cubit ropes to remeasure the area of their fields. The ropes were knotted at every cubit and carried by teams of men, known as rope stretchers.

Egyptian surveyors with measuring ropes

EGYPTIAN CUBIT

The ancient Egyptians measured length using a unit called a cubit. This was the distance between a person's elbow and the tip of his or her outstretched middle finger—about 19.5 inches (50 centimeters). To avoid confusion, a standard cubit was made from black granite. Called the Royal Cubit, it measured 20.4 inches (52.4 centimeters). All other cubit sticks were measured against this one.

FEET FIRST

The human foot was first used as a unit of measurement in about 1500 B.C. by the Babylonians. The Babylonians' foot was about 13 inches (33 centimeters) long. The Romans used a foot of about 11.7 inches (30 centimeters). The earlier Greek foot measured the same but was based on the width of 16 fingers. The modern foot is 12 inches.

The measuring rod in this mosaic shows the length of a Roman foot.

MILES AND MILES

The word *mile* comes from the Latin *mille*, which means "a thousand." A Roman mile was 1,000 paces long (about 1,591 yards/1,460 meters), each pace being two of a Roman soldier's strides. A modern mile measures 1,760 yards, which is a little more than 1,609 meters.

INCH BY INCH

The Romans divided the foot into 12 parts called *unciae*. An *uncia* was the width of a man's thumb. *Uncia* gives us the word *inch*. There are 12 inches to a foot. *Uncia* also gives us the word *ounce*, which was once a twelfth of a pound. There are now 16 ounces to a pound.

A ruler divided into 12 inches

A Roman milestone. Milestones were placed at one-mile intervals along every Roman road.

King Henry I was responsible for fixing the length of the first standard yard.

IN FACT...

The Egyptians, Greeks, and Romans all used hands and fingers as units of measurement. The Egyptians used the digit (the width of a man's middle finger) and the palm (the distance across his palm). The Greeks' basic unit of length was the finger. One finger equaled .8 inch (19.3 millimeters). The hand is still used today to measure the height of horses. One hand equals 4 inches (10 centimeters).

BY THE YARD

The yard was a unit of measurement invented by traders to measure cloth. Each yard was a length of cloth stretched between chin and fingertip. In 12th-century England, the yard was fixed as the distance between the end of King Henry I's nose and the tip of his middle finger. In 1305 the yard was reset at three feet and remains so today.

Unusual units

Over the centuries, hundreds of different units have been devised to measure almost everything—from stalks of corn to molecules. Many units were invented to weigh and measure everyday things. Others have one particular function—to measure gold or gemstones, for example.

Megaliths were gigantic stones used to build ancient tombs such as this one in Cornwall, England.

MEGALITHIC YARD

Some 5,500 years ago, people in western Europe constructed tombs from large upright stone slabs, covered with a horizontal capstone. They took one of their basic units of length from these huge stones, or megaliths. As a result, this measurement is known as a megalithic yard. The megalithic yard has been calculated at 32 inches (82.9 centimeters).

LAND MEASURES

The word *acre* comes from the Greek word for "field," *agros.* Originally, an acre was the amount of land a team of oxen could plow in one day. It was later fixed at an area 40 rods long by 4 rods wide, a rod being 5.5 yards (5 meters) long. A furlong, now used to measure horse races, was originally the length of a furrow plowed by a team of oxen without stopping. Today, a furlong is fixed at 220 yards (201 meters).

In the Middle Ages, farmers used teams of oxen or horses to plow their fields.

PRECIOUS CARATS

Gold, diamonds, and pearls have their own unit of weight, called the carat. One carat equals 200 milligrams. The carat is also a measure of purity for gold. Only 24-carat gold is pure; lower-carat gold is mixed with cheaper metals. The word *carat* comes from the Greek word *keration*, which means "carob bean." The carob bean was once used as a natural unit of weight.

KNOTS AND NAUTICAL MILES

Distance and speed at sea are measured in nautical miles and knots. A nautical mile is 1.15 land miles (1.85 kilometers). A knot is the speed of one nautical mile per hour. In the past a ship's speed was calculated by dropping a rope, knotted at regular intervals, over the side. By counting the number of knots let out over a fixed period of 28 seconds, sailors could measure how fast the ship was traveling. This is how the knot got its name.

A gem-encrusted crown from the British crown jewels

SIGNPOST

The Greek currency, the drachma, began life as a measurement of weight. The word *drachma* means "handful." In ancient Greece a drachma, or handful, was made up of six tiny metal weights, called *obols*, which would fit into a hand.

Drachmas are still used as the currency of Greece. The coins show great Greek figures, such as the philosopher Aristotle.

BREAKTHROUGH

Some units are used to measure very small amounts. Molecules are so tiny that scientists cannot measure them individually. Instead, they measure molecules by weighing them in groups. The unit of measurement for molecules is called a mole. A mole of any substance is made up of 6×10^{23} (this means 6 followed by 23 zeros) molecules!

SYSTEMS AND STANDARDS

As trade developed, more accurate, standardized systems of weights and measures were needed. Early systems varied so much from place to place, and from item to item, that weighing and measuring became rather haphazard. A yard of cloth in one place might be several inches shorter than in another. Today, units of measurement are standardized all over the world, to a very high level of accuracy.

A medieval French fair, held every year for two weeks in June

CHARLEMAGNE'S FOOT

In the 9th century, the emperor Charlemagne tried to introduce a standard foot by fixing the measurement as the length of his own foot. Other attempts to standardize measurement were made at the great European trade fairs of the 12th and 13th centuries. All the merchants at the fair were forced to use identical weights and measures. The Keeper of the Fair checked these against his own master set.

IMPERIAL MEASURES

The imperial system uses units such as pounds and ounces, yards, feet and inches, pints and gallons. It was established in Britain in the 1300s and soon spread all over the world. Today, the imperial system is mainly used in the United States.

The standard weights and measures, including a yardstick, used in Elizabethan times (1558-1603) in Great Britain

1 IN FACT...

A set of standard measures, originally based on the length of King Henry I's arm (see page 9), was kept in the Houses of Parliament in London. These could be copied and used all over the country. In 1834 both the building and the standards went up in flames. A new yard bar had to be made. It was kept in a fireproof box, bricked up in the wall of the new House of Commons. Every 20 years, it was taken out in order to check the standard copies.

Gold ingots, each weighing 400 troy ounces

TROY WEIGHTS

The troy system of weights and measures was used in the Middle Ages for weighing gold, other precious metals, and medicines. This system was invented in the French town of Troyes. The troy pound contains only 12 ounces rather than the 16 ounces in an imperial pound. It is still used today for weighing precious metals.

INTERNATIONAL UNITS

In 1960 a new, international system of weights and measures was formed for use mainly by scientists all over the world. It is known as the SI (*Système International*). It uses six basic units of measurement—the meter (length), the kilogram (weight), the second (time), the ampere (electrical current), the kelvin (temperature), and the candela (brightness of light).

BREAKTHROUGH

Today, most countries use the metric system, invented in France in the 18th century. The main units of measurement are meters and centimeters, grams, kilograms, and liters. For the first time, these units were not based on the body. The meter was originally fixed as the distance between the North Pole and the equator, divided by 10 million. The modern meter is fixed in a laboratory by measuring how far light travels in a set amount of time (one 299,792,458th of a second).

WHAT A WEIGHT

Objects have weight because of the force of gravity pulling them down toward the earth. When you weigh something, you are actually measuring the effect of gravity on the object's mass. Mass is how much material something contains. An object's mass never changes. But an object's weight can change. On the moon, for example, an object would weigh six times less than on earth, because the moon's gravity is much weaker than that of earth.

Astronauts have to cope with eating, drinking, getting dressed, and going to the bathroom in a state of zero gravity!

1 N FACT...

In the Middle Ages, women who were suspected of witchcraft were forced to undergo trial by weight. They were thrown into a river or pond. If they floated, they were judged to be guilty. If they sank, they were innocent—but had probably drowned.

WEIGHT IN SPACE

In space, there is no gravity to pull people down. This is why astronauts float around their spacecraft in a state of weightlessness. To help them get used to this strange feeling, astronauts train underwater and in specially adapted aircraft.

UNITS OF WEIGHT

Weight is measured in grams and kilograms (metric) and in ounces and pounds (imperial). One thousand kilograms equals about one ton. The universal standard for the kilogram is a metal cylinder kept in the International Bureau of Weights and Measures in France. Copies of the standard kilogram, called witnesses, are kept in various places all over the world.

Weighing scales showing both grams and kilograms and ounces and pounds

MEASURING VOLUME

Volume measures the amount of space a liquid or object takes up. In the metric system, volume is measured in cubic meters or in liters. The mass of one liter of water equals one kilogram. In the imperial system, volume is measured in fluid ounces, pints, and gallons.

WORTH YOUR WEIGHT IN GOLD

The Mogul emperor Jahangir ruled India from 1605 to 1627. According to eyewitness reports, the emperor's birthday was celebrated in grand style. Jahangir was weighed on a set of golden scales against a pile of gold and precious stones. The emperor's weight in gold and gems was then shared among the people.

Laboratory flasks. The amount of space a liquid takes up is called its volume.

Jahangir also weighed his son Prince Khurram against a pile of gold and silver on the prince's birthday.

SIGNPOST

Before coins were invented, people used gold, silver, and copper as money. The weight of the metal was the important thing. Its size and shape did not matter. The metal lumps were stamped with symbols stating their weight and value. The symbol also identified the person who was guaranteeing that the metal weighed the correct amount.

LIGHT AS A FEATHER

The ancient Egyptians believed that a virtuous person would enter the Next World when he or she died. To test this, the dead person's heart was weighed against the Feather of Truth. If the person had led a sinful life, his or her heart would be heavier than the feather and the person would be fed to a monster. If the person had led a good life, his or her heart would balance the feather and he or she would be allowed into the Next World.

Anubis, the ancient Egyptian god of the dead, weighs the heart of a dead person against the Feather of Truth.

KEEPING CHECK

In ancient Greek times special officials were appointed to check the weights and measures used by market traders to make sure that no one was being cheated. The officials checked the traders' weights against their own standard set. The officials were called *metronomoi*, after the Greek word *metron*, meaning "measure."

ROYAL AND COMMON

In ancient times two different measures of weight were used in the Middle East. They were called the "royal" and the "common" measures. The royal measure was the bigger of the two. Taxes paid to the king were calculated using the royal measure. But when the king had to pay money, he used the common measure. So the king always made a profit!

Ancient Greek weights were stamped with turtles, dolphins, and other emblems to show how much each piece of metal was worth.

IN FACT...

The gigantic blue whale is the biggest animal that has ever lived— bigger even than the dinosaurs. Adult blue whales can weigh up to 165 tons (150 metric tons). A blue whale's tongue alone weighs 3.3 tons (3 metric tons), as much as a rhinoceros!

A blue whale

HEAVYWEIGHT EARTH

Earth has a mass of 6.6 octillion tons (6 octillion metric tons). (An octillion is 1 followed by 27 zeros.) This was first calculated in 1174 by Scottish scientist Nevil Maskelyne. Water covers more than two-thirds of the earth's surface, and its total weight is 1.55 quintillion tons (1.41 quintillion metric tons), or 0.024 percent of the earth's weight. (A quintillion is 1 followed by 18 zeros.)

WEIGHT LIFTING

Weight lifters train hard and follow a special diet to build up their muscle power. The strongest weight lifters can lift the equivalent of three times their own body weight above their heads.

A mixture of strength and skill allows weight lifters to lift very heavy weights.

BREAKTHROUGH

The Greek mathematician Archimedes (287-212 B.C.) is best known for a law called Archimedes' Principle. This law states that when an object is put into water, it is pushed upward by a force equal to the weight of the water it has displaced. This force is called upthrust. If the upthrust is equal to the object's weight, the object will float. Archimedes is said to have made his discovery as he got into his bath.

The brilliant Greek mathematician and engineer Archimedes. He is said to have shouted "Eureka!" ("I have found it!") when he made his great discovery in the bath.

LONG WAYS AND LIGHT-YEARS

Lengths and distances are usually measured from one point to another along a straight line. Length is used to measure the size of an object itself. Distance is the measurement of the space between two places or objects. Speeds are calculated according to the distance covered in a fixed period of time.

The stars of the Southern Cross are 500 light years away

IN FACT...
The closest star to the earth is the sun, 93 million miles (150 million kilometers) away. The next nearest, Proxima Centauri, is 4.25 light-years away.

AT LENGTH
The main units used to measure length and distance are centimeters, meters, and kilometers (metric) and inches, feet, and miles (imperial). Distances at sea are measured in nautical miles (see page 11). The area of an object is its length multiplied by its width. Area is measured in square units, such as square meters, square feet, or square miles. Speed is measured in kilometers or miles per hour.

LIGHT-YEARS
Distances in outer space are so vast that astronomers use a special unit, called a light-year, to measure them. A light-year is the distance that light travels in a year—5.9 trillion miles (9.5 trillion kilometers). A unit called an astronomical unit (AU) is also used in space. An AU is the average distance between the earth and the sun, equal to about 93 million miles (150 million kilometers).

The Concorde is a supersonic aircraft. This means that it flies faster than the speed of sound.

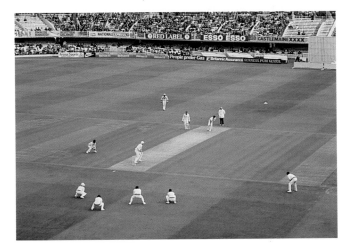
A cricket pitch measures 22 yards (20.2 meters).

SPEED OF SOUND

One of the first people to calculate how fast sound travels was a man named William Derham, in 1708. He stood on top of a church tower while a cannon was fired 12 miles (19 kilometers) away. By timing the interval between the cannon's flash and its boom, he calculated the speed of sound. His answer was not far off the modern figure, which scientists have calculated as 1,125 feet (343 meters) per second, at a temperature of 68°F (20°C). Sound travels faster in higher temperatures. It also travels much faster (about five times faster) through water.

IN CHAINS

The chain is an ancient unit of measurement, 22 yards (20.2 meters) long. It has been used by land surveyors for more than 350 years. Originally a real chain made up of 100 links was used. Today, the chain is mainly used as the standard measure for cricket pitches, or playing fields.

SIGNPOST

Being able to calculate distances is essential in finding your position in navigation. Early sailors navigated by the sun and the stars using a sextant. Later, they used a system of latitude (how far north or south of the equator they were) and longitude (how far east or west of the Greenwich Meridian they were). Modern navigators have sophisticated equipment to help them, including radar, satellites, and detailed maps, called charts.

The sextant used by the famous explorer David Livingstone

Taking the Temperature

You can tell whether the weather is hot or cold simply by standing outside. But to judge exactly how hot or cold, you need a thermometer. Temperature is a measure of how hot an object is. The hotter it is, the higher its temperature. Very cold objects have negative temperatures, below zero.

The world's largest thermometer, in California. It shows air temperatures in the range of 20°F to 140°F (-7°C to 60°C).

FAHRENHEIT SCALE

Temperature is measured in degrees, using two scales—Fahrenheit and centigrade, or Celsius. The Fahrenheit scale was invented in 1718 by the German instrument maker Gabriel Daniel Fahrenheit (1686-1736). First he fixed 0°F by using the coldest substance he knew, a freezing mixture of ice and salt. The beginning and end of the scale were marked by the freezing point of water (32°F) and its boiling point (212°F).

CENTIGRADE SCALE

Most countries today use the centigrade scale, invented by the Swedish astronomer Anders Celsius in 1742. His scale was divided into 100 degrees, with water freezing at 0°C and boiling at 100°C. Originally, Celsius had water boiling at 0°C and freezing at 100°C, but this was reversed after his death!

SIGNPOST

Use these simple equations to convert from Fahrenheit to Celsius and the other way around:

°F = (°C x 1.8) + 32

°C = (°F – 32) ÷ 1.8

A digital thermometer showing the temperature of melting ice

BREAKTHROUGH

The first sealed thermometer was invented in 1641 by Grand Duke Ferdinand of Tuscany, Italy. It was much more accurate than earlier versions. Today, several different types of thermometers are used. Glass thermometers contain a column of mercury or alcohol that rises and falls as the temperature increases and decreases. Modern digital thermometers show the temperature as numbers on a screen.

Red-hot lava flows from Mount Etna, Sicily, in 1992. Trenches were dug to divert the flow so that the lava could cool and cause less damage.

HOT STUFF

An instrument called a pyrometer is used to measure very high temperatures, such as the red-hot lava erupting from a volcano. Lava can reach temperatures of about 1,100°F (600°C). The word *pyrometer* means "fire measurer." A pyrometer works by matching the changing color of an object as it gets hotter with the changing color of an electric filament inside the instrument.

BODY TEMPERATURE

The temperature inside your body usually stays at a constant 98.6°F (37°C). This is the temperature at which your body works best. Sweating helps to cool you down; shivering helps to warm you up. It can be dangerous if your body temperature falls below 77°F (25°C) or rises above 106°F (41°C).

This girl uses a thermochromic, or strip, thermometer. The patches on the thermometer change color to indicate body temperature.

21

Lord Kelvin, the Scottish physicist and mathematician. He discovered the existence of absolute zero in 1848, while he was professor of natural philosophy at Glasgow University.

ABSOLUTE ZERO

The lowest temperature possible is -456°F (-273°C), known as absolute zero. Temperature is the measure of how fast the molecules in a substance are moving. The faster they move, the higher the temperature. Absolute zero is the temperature at which all such movement stops. Absolute zero was discovered by the Scottish physicist Lord Kelvin (1824-1907). It is also written as 0°K (Kelvin). Scientists have been able to cool substances almost to absolute zero, but have never reached the exact temperature.

SIZZLING SUN

The temperature at the center of the sun is 27 million °F (15 million °C). If a pinhead were this hot, it would set fire to everything for 62 miles (100 kilometers) around it! The temperature at the center of the earth is about 8,100°F (4,500°C).

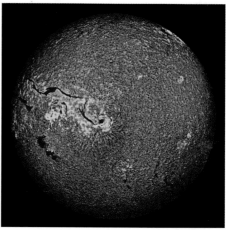

The red-hot sun

IN FACT...

The hottest place on earth is Dallol, Ethiopia, where the average annual temperature in the shade is 94°F (34.4°C). The coldest place is the Pole of Inaccessibility in Antarctica. Here, the average yearly temperature is -72°F (-58°C).

Death Valley, California (above), is one of the hottest, driest places on earth. The Antarctic (left) is the world's coldest, windiest place.

PLANET CLIMATES

Earth is the only planet known to have just the right temperature to support life. The other planets in our solar system are too hot or too cold. Venus is the hottest planet, with a scorching surface temperature of 864°F (462°C). The surface of Pluto, by contrast, is a freezing -382°F (-230°C).

WEATHER MEASURES

Apart from thermometers, many other instruments are used to measure the weather. These measurements allow meteorologists to make weather forecasts. Barometers measure air pressure in units called millibars. Rain gauges measure inches or millimeters of rain. Anemometers and weather vanes measure wind speed (in miles or kilometers per hour) and direction (north, south, east, or west). Humidity is measured with a thermometer or with an instrument called a hygrometer. Weather satellites are also used to provide information.

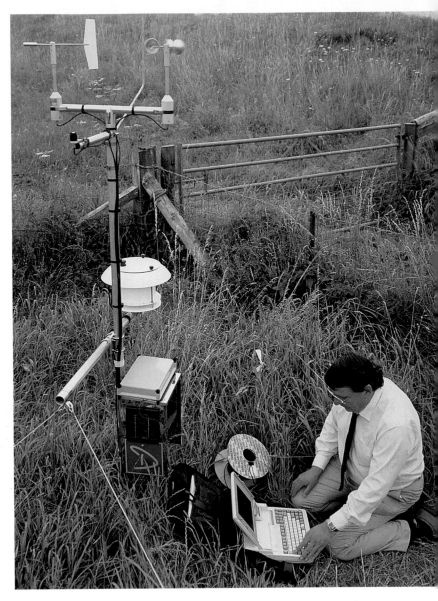

A weather-monitoring station. It has instruments for measuring wind speed and direction, temperature, humidity, and sunshine.

Follow the fish to find out which way the wind is blowing!

TRACKING TORNADOES

Tornadoes are violent, twisting storms that cause great damage as they race across the land. Tornadoes are difficult to predict because they happen so quickly. But a special scale, the Torro tornado intensity scale, is used to measure them on a scale of 1 to 12. Torro force 1 is mild and can uproot small trees, for example. Torro force 12 is a supertornado, likely to damage even the strongest, steel-reinforced buildings.

In the Balance

Early people had little need for accurate weights or weighing machines. If they wanted something, they simply bartered (exchanged something similar) for it. With the growth of trade between different countries, however, more precise ways of calculating weights and measures soon became essential.

SCALES FOR GOLD

In ancient Egypt gold was so precious that, in about 3500 B.C., the Egyptians invented scales to weigh it accurately. The scales were a simple beam balance, with a pan hung on either end of a beam. The gold was put in one pan and balanced against a set of weights in the other.

WEIGHING TOOLS

Metal was so precious in ancient Egypt that metal tools were carefully weighed before being given out to workers. They were weighed again when they were returned. This was done to make sure that the workers did not shave off bits of the valuable metal.

Ancient Egyptian metal workers weigh gold on simple beam-balance scales.

A medieval beam balance is used to weigh out quantities of dried meat.

ROMAN STEELYARD

The Romans invented the steelyard in about 200 B.C. A steelyard has one arm longer than the other. An object is hung from the shorter arm and a weight is moved along the longer arm until the arm balances. The position of the weight when the arm is balanced shows the weight of the object.

ON A KNIFE EDGE

In the 18th century, weighing machines became much more accurate with the invention of the knife-edge balance. This is a very accurate beam balance. The beam balances on a knife edge so that both arms are exactly equal in length.

EARLY WEIGHTS

Early weights came in many forms. The Egyptians used weights in the shape of bulls' heads to weigh gold. The Ashanti of Africa made brass weights in the shapes of animals and fish. To make the weights lighter, they simply chopped off a leg or a tail!

A Roman steelyard

Weights used in the 1600s by jewelers in France for weighing out their materials.

IN FACT...

In ancient China sound was used to measure volumes of grain and wine. The shape and fullness of a container governed what sound it would make if struck with a stick. If a filled container sounded at the right pitch, then it held the correct amount.

An Ashanti brass weight in the shape of an elephant. It was used for weighing gold dust.

SIGNPOST

A stone is a unit of weight used in the imperial system. Originally, it was based on the weight of an actual stone and was used to weigh wool. Today, a stone equals 14 pounds (6.3 kilograms) and is mainly used in Britain, for weighing humans.

MODERN MEASURES

Many modern weighing and measuring machines are amazingly precise instruments, designed to meet the demands of trade, science, medicine, and technology. There are now accurate instruments to measure anything from the weight of a gas to the temperature of a star in a far-distant galaxy.

MICROBALANCES

The most sophisticated scales are the microbalances used in scientific research. They are used mainly to weigh small amounts of gases. The most sensitive microbalances can measure substances to a millionth of a gram—less than the weight of an eyelash!

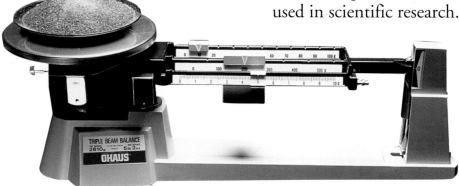

A triple-beam balance used in a chemistry laboratory. Each beam has a different scale (tens of grams, hundreds of grams, and tenths of a gram) for very accurate measuring.

ECHO SOUNDING

Instead of dropping a weighted rope over the side of a ship, as sailors did in the past, there is now sophisticated equipment for measuring the depth of the seas. Scientists, sailors, and fishermen use devices called echo sounders, which send out beams of sound. On-board computers record the returning echoes and use them to calculate how deep the water is.

Fishermen use echo sounders to locate large schools of fish, as shown on this computer screen.

BREAKTHROUGH

A highly accurate electronic weighing machine for weighing very small amounts

Electronic weighing machines were developed in the 20th century and are much more accurate than the mechanical balances of the past. Electronic scales can be used not only to find an object's total weight but also to give other information such as average weight and moisture content.

HIGH-TECH RULER

The laser interferometer is the latest and most accurate ruler there is. It is used for the precise calibration (measuring out) of tapes and other instruments used for measuring length.

LASERS AND TELESCOPES

Lasers, satellites, and radio telescopes are used by astronomers to measure the vast distances in space. Radio telescopes pick up signals from stars hundreds of light-years away. A laser is an intense beam of light that can travel great distances at a constant speed. Lasers have been used to calculate distances in space more accurately than ever before.

IN FACT...

Instead of using parts of the body, standards of length are now set in laboratories using sophisticated machines. The standards are fixed using the speed of light so that they are extremely precise.

Scientists use lasers to measure vast distances in space. In this picture, the horizontal beam is used to measure the distance between the earth and the moon. It is accurate to within 1 inch (3 centimeters)! The vertical beam measures the distance from the earth to satellites orbiting in space.

TIMELINE

B.C.

— *c.* **3500** The Sumerians invent the earliest types of beam balances

— *c.* **3500** The megalithic yard is used in Europe

— *c.* **3500** First scales invented in Egypt

— *c.* **3000** The Egyptians invent the cubit

— *c.* **2600** The Babylonians use the *mina*, the earliest known type of weight [it equaled about 34 ounces (970 grams)]

— *c.* **1500** The Babylonians are the first to use the human foot as a unit of measurement

— **287-212** Life of Archimedes, Greek mathematician

— *c.* **200** The Romans invent the steelyard

A.D.

— **900s** Emperor Charlemagne fixes the foot as the length of his own foot

— **1100s** The yard is fixed as the distance between the end of King Henry I's nose and the tip of his middle finger

— **1300s** The imperial system is established in Britain

— **1305** The yard is reset at 3 feet (.9 meter)

— **1638** Micrometer invented for measuring in astronomy

— **1641** First sealed thermometer invented by Grand Duke Ferdinand of Tuscany

— **1700s** Knife-edge balance invented

— **1708** William Derham calculates the speed of sound

— **1718** Fahrenheit temperature scale invented by Gabriel Daniel Fahrenheit

— **1742** Centigrade temperature scale invented by Anders Celsius

— **1793** Metric system invented in France

— **1824-1907** Life of Lord Kelvin, Scottish physicist

— **1960** The SI (*Système International*) is introduced

— **1973** First electronic micrometer invented

— **1976** NASA launches LAGEOS (the laser geodynamics satellite) for measuring vast distances accurately

— **1983** The International Conference on Weights and Measures redefines the length of the meter

GLOSSARY

Altitude The height of an object above sea level.

Calibration The very precise measuring out and marking of tape measures and other measuring instruments.

Carat A unit of weight used to measure gold, diamonds, and pearls. One carat equals 200 milligrams. Used as a measure of purity for gold—pure gold is 24 carats.

Furlong A unit of distance measuring 220 yards (201.1 meters). Originally the length of a furrow plowed by a team of oxen without stopping to rest.

Gravity An invisible force that pulls things toward the center of the earth or toward each other.

Greenwich Meridian An imaginary line running through Greenwich, England, that marks 0° longitude.

Imperial system A system of weights and measures that uses units such as pounds, ounces, feet, inches, miles, and gallons.

Light-year A light-year is the distance that light travels in a year—5.9 trillion miles (9.5 trillion kilometers); used to measure distances in space.

Mass An object's mass is the amount of material that object contains. When you weigh an object, you are actually measuring the effect of gravity on its mass.

Mercury A white, normally liquid metal used in thermometers and barometers.

Metric system A system of weights and measures that uses units such as meters, centimeters, kilograms, liters, and kilometers.

Microwave Very short energy waves, used for cooking food, making international telephone calls, and measuring.

Molecule A tiny particle made up of even tinier particles, called atoms. You, and everything around you, are made of molecules.

Navigation Plotting the position of and course taken by a ship or an aircraft.

Satellite An object that orbits the earth. Satellites are used for communication, map making, forecasting the weather, and measuring vast distances.

SI The *Système International* of weights and measures. This was devised in 1960 as a standard system for use by scientists around the world.

Supersonic Faster than the speed of sound.

Temperature The measure of how hot or cold something is.

Volume The amount of space that something takes up. It is often used as a measure of liquids.

Weight The measure of how heavy something is. Weight is the effect of gravity on an object's mass.

INDEX